Steck-Vaughn
Strength in Numbers™
Algebra
Level 5

Acknowledgments

Editorial Director: Diane Schnell
Supervising Editor: Donna Montgomery
Associate Director of Design: Cynthia Ellis
Design Manager: Deborah Diver
Production Manager: Mychael Ferris-Pacheco
Production Coordinator: Susan Fogarasi
Editorial Development: Words and Numbers
Production Services: Mazer Corporation

ISBN: 0-7398-6248-0

Strength in Numbers™ is a trademark of Steck-Vaughn Company.

Copyright © 2003 Steck-Vaughn Company

STECK-VAUGHN
A Harcourt Company

www.steck-vaughn.com

Contents

Expressions and Equations

Up, up, and away!

A hot air balloon lifts off when the air inside the balloon is warmer than the outside air.

You only have to be 14 years old to get a student license to fly a hot air balloon. But you have to know a lot of math. For a balloon to carry two people, the bag must hold about 60,000 cubic feet of air. How do you think you would find the amount of air a balloon might hold?

Numerical Expressions

An **expression** is a mathematical phrase that does not have an equal sign. It is used to combine numbers and operation signs.

Examples: $4 + 6$ $9 - 5$ 3×3 $28 \div 2$

Write an expression to model each situation. Use numbers and operation signs. Then find the value of each expression.

Model 1

Heather is 2 years older than Jorge. Jorge is 13 years old. Write an expression to show how to find Heather's age.

What do you know?   *Jorge's age is 13.*
                        Heather is 2 years older than Jorge.

The clue in the problem is the word *older.* It tells you to add, because Heather's age is more than Jorge's age.

Jorge's age	**plus**	**2 years**
_____	_____	_____

The value of the expression for Heather's age, *13 + 2* is _____.

Model 2

Alan has 4 fewer comic books than Ellen, and Miguel has 1 fewer comic book than Alan. Ellen has 16 comic books. Write an expression to show how many comic books Miguel has.

Ellen's books	**minus**	**4 books**	**minus**	**1 book**
_____	_____	_____	_____	_____

The value of the expression is _____. Miguel has _____ comic books.

Explain how you can work backward to check that the answer of 11 comic books is reasonable.

_____

_____

_____

Practice

Write an expression to model each situation. Use numbers and operation signs. Then find the value of each expression.

1. Sandra ate 11 grapes. Then she ate 5 more grapes.

   Expression: ___11 + 5___

   Value: ___16___

2. Calvin had 18 baseball cards. He sold 11 of the cards.

   Expression: _____

   Value: _____

3. Howard picked 14 apples. He gave away 5 apples.

   Expression: _____

   Value: _____

4. Mia is 12 years older than her cousin. Her cousin is 8.

   Expression: _____

   Value: _____

Some expressions have more than one step. Write an expression to model each situation. Then find the value of each expression.

5. Robert had 21 stamps. He gave away 4 of them. He bought 7 more stamps.

   Expression: _____

   Value: _____

6. Tanya was 11 years old 3 years ago. Elena is 2 years older than Tanya. Find Elena's age now.

   Expression: _____

   Value: _____

7. Will gave 8 books to a book drive. Then he gave 2 more fiction books and 5 nonfiction books.

   Expression: _____

   Value: _____

8. Latisha had 14 baseball cards. She traded 4 of her cards for 2 of Kim's cards.

   Expression: _____

   Value: _____

 Write your own problem to match the expression 10 + 14. Find the value of the expression.

_____

_____

_____

Lesson 2 — Variables and Expressions

Sometimes an expression has a **variable**, which is a letter that stands for a number. You can use a variable when an exact amount is not known.

Examples: $4 + t$ $n - 2$ $3 \times p$ $n \div 5$

A variable can be used in an expression to model a situation.

Model 1

Raul collected comic strips and kept them in a shoebox. He had 21 of them. His sister put more comic strips into the box. **Write an expression with a variable to model the situation.**

21 comic strips	plus	more comic strips
_____	_____	_____

You can substitute a number for a variable to find the value of an expression.

Model 2

If $p = 88$, what is $p - 30$?

Substitute the value for the variable in the expression.

$p - 30 =$

$88 - 30 =$ _____

Because $88 - 30 = 58$, the value of the expression is 58.

 How can you find the value of an expression with a variable? Write a sample problem to show the process.

_____

_____

_____

_____

_____

_____

_____

Practice

Write an expression with a variable. Explain what the variable stands for.

1. Liz had 7 baseball cards. She got a few more at the store.

   Expression: _____ 7 + n _____

   The variable stands for

   _____ **more cards** _____.

2. Samuel had 30 pencils. He sold many of them.

   Expression: _____

   The variable stands for

   _____.

3. There were 8 plates on the shelf. Cal took some of them off the shelf.

   Expression: _____

   The variable stands for

   _____.

4. A number of cups were on the table. Kelly put 6 of them away.

   Expression: _____

   The variable stands for

   _____.

Substitute the given value for each variable. Write the value of each expression.

5. If $n = 10$, what is $5 + n$?

   The value is _____ 15 _____.

6. If $p = 12$, what is $p - 4$?

   The value is _____.

7. If $m = 22$, what is $24 - m$?

   The value is _____.

8. If $p = 25$, what is $100 - p$?

   The value is _____.

Use each value for n to complete the tables.

9.

n	$n + 4$
8	12
4	
12	
7	

10.

n	$18 - n$
6	
9	
11	
3	

11.

n	$n - 5$
25	
10	
17	
14	

Write your own problem to match the expression $19 + n$. Substitute a value for the variable and find the value of the expression.

_____

_____

Addition and Subtraction Equations

An **equation** is a mathematical sentence that has an equal sign. The value left of the equal sign is the same as the value right of the equal sign.

Examples:

$13 - 8 = 5$	$4 + 2 + 6 = 12$	$7 + 9 = 9 + 7$
$5 = 5$	$12 = 12$	$16 = 16$

Write an equation to model the situation.

Model 1 ▶ Jeff's dog weighed 6 pounds last year. It gained 8 pounds during the year. Now Jeff's dog weighs 14 pounds.

weight last year (lb)	plus	weight gained (lb)	is	current weight (lb)
_____	+	_____	=	_____

Some equations contain variables. **Example:** $n + 11 = 15$

Write an equation with a variable to model the situation.

Model 2 ▶ Erin rode her horse for 7 miles of a 12-mile trail. How many more miles does she have to ride to complete the trail?

Let *y* represent the number of miles left to ride. Use a model to represent the situation.

12-mile trail	
7 miles	*y* miles

miles Erin rode	plus	miles left to ride	is	length of trail (mi)
_____	+	_____	=	_____

Use mental math.

How many miles does Erin still have to ride? _____

What is the value of *y*? $y =$ _____

Explain how you can check your answer.

_____

_____

Practice

Write an equation to model each situation.

1. Chun sold 35 tickets to the school carnival. Megan sold 46 tickets. Together they sold a total of 81 tickets.

   $35 + 46 = 81$

2. Jessie had 140 stamps in his collection. He sold 31 stamps. He has 109 stamps left.

3. A sweatshirt cost $31.95 before the sale. It was reduced $5.00 to make it $26.95 during the sale.

4. Students in Jada's class made 15 posters, 6 posters, and 9 posters for the class election. They made 30 posters in all.

**Write an equation with variable *n* to model each situation.
Then find the value of *n*.**

5. Sal planted 12 flowers. He needs to plant 24 flowers to fill the flowerbed. How many more flowers does he need to plant?

   $12 + n = 24$

   $n = 12$

6. Sarah did 42 hours of volunteer work at the Senior Center. She will earn a badge for 50 hours. How many more hours does Sarah need to work to earn the badge?

7. After selling 3 handmade kites at the Craft Fair, Kim had 9 kites left. How many kites did Kim start with?

8. The high temperature one day was 72°F, and the low temperature was 49°F. What was the difference between the temperatures?

   How are equations and expressions alike? How are they different?

_____

_____

_____

Lesson 4

Multiplication and Division Equations

Sometimes multiplication or division is needed in an equation to model a situation. For example, to find the number of eggs in four dozen, you may write $4 \times 12 = 48$.

Examples of multiplication and division equations:

$$4 \times 7 = 28 \qquad 10 \div 2 = 5$$

Write an equation to model the situation.

Model 1

Ellen used 72 beads to make 8 necklaces. She put 9 beads on each necklace.

total number of beads	divided by	number of necklaces	is	number of beads on each necklace
_____	÷	_____	=	_____

Some multiplication and division equations contain variables.

Examples: $12 \times n = 144 \qquad y \div 5 = 9$

Write an equation with a variable to model the situation.

Model 2

Members of the Rodeo Club paid $4 for each ticket to the rodeo. They paid $40 in all. How many tickets did they buy? **Use a model to represent the situation. Let *n* represent the number of tickets bought.**

price per ticket	times	number of tickets	is	total amount paid
_____	×	_____	=	_____

Use mental math. How many tickets did they buy? _____

What is the value of *n*? _____

How can you check your answer?

_____

_____

Practice

Write an equation to model each situation.

1. Students in the Book Club donated a total of 21 books to 3 day care centers. Each center received 7 books.

   21 ÷ 3 = 7

2. Juan swam 8 laps of the swimming pool every day. In 4 days, he swam 32 laps.

3. A total of 20 people are on the roller coaster. There are 10 cars on the roller coaster with 2 people in each car.

4. Latoya watches 5 hours of television every week. In 6 weeks, she watches 30 hours of television.

Write an equation with variable _n_ to model each situation. Then find the value of _n_.

5. Elisa paid $30 for 5 tropical fish. Each fish costs the same amount. What did Elisa pay for each fish?

6. Julia filled 16 glasses with juice. Each glass held 8 ounces. How many ounces of juice did Julia use?

7. Ben sent 8 postcards to his friends. Each stamp costs $0.21. How much did Ben spend on stamps?

8. Pat played his favorite song over and over again for 24 minutes. The song is 3 minutes long. How many times did Pat play the song?

Write your own problem to match the equation 36 ÷ _n_ = 9. Identify what the variable represents.

_____

_____

Equations and Substitution

You can use **substitution** to solve an equation. This means to replace the variable with a specific amount. The value of the variable that makes the equation true is the **solution** of the equation.

Substitute a value for *s* in the equation. Then check your solution.

Model 1

$56 \div s = 8$

Think: 56 divided by what number is equal to 8? _____

Substitute 7 in place of *s*.

$56 \div$ _____ $= 8$

$8 = 8$

The solution of the equation is 7, because when $s =$ _____, both sides of the equation are equal.

Write an equation with variable *n* to model the situation.

Model 2

Jeff had $15 in his bank account on June 1. By June 30, he had $45 in the account. If Jeff did not take any money out of his account, how much did Jeff add to the account in June?

amount on June 1	plus	amount added	is	amount on June 30
_____	+	_____	=	_____

Estimate a solution to substitute. Then check.

What if you estimate $n = \$30$?

$\$15 +$ _____ $= \$45$

$\$45 = \$45$, so $n = \$30$ is the solution.

 What if your first guess does not solve a problem? Explain how a wrong guess can help you find a solution to a problem.

_____

_____

Practice

Use substitution to solve each equation. Then check your solutions.

1. $n - 31 = 10$

$n = 41$

$41 - 31 = 10$

$10 = 10$

2. $7 \times s = 49$

3. $85 + 6 = y$

4. $24 \div y = 6$

5. $150 + n = 200$

6. $s \times 10 = 100$

Write and solve an equation to model each situation. Then check your solution.

7. Pedro used 2 yards of fabric to make one puppet. How many puppets did he make if he used 6 yards of fabric? Let p = number of puppets.

$2 \times p = 6$

$p = 3$

$2 \times 3 = 6$

$6 = 6$

8. Anna worked 2 hours in the morning. By the end of the afternoon she had worked a total of 5.5 hours. How many hours did Anna work in the afternoon? Let h = hours of work.

9. Carl earned $35 washing cars. He charged $5 to wash each car. How many cars did he wash? Let c = number of cars washed.

10. Tricia paid for a roll of film with a $10 bill. She got $2.25 back in change. What was the price of the roll of film? Let p = price of film.

  Is $y = 70$ a solution of the equation $11 + y = 88$? How do you know?

_____

_____

6 Balanced Equations

An equation is like a balance scale. Both sides must have the same value.
An equation is **balanced** if both sides are equal.

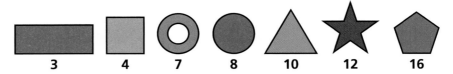

| 3 | 4 | 7 | 8 | 10 | 12 | 16 |

Select the correct shapes to balance the scale. Draw the shapes that will make the scale balanced.

 Model 1

What value is needed to balance the scale?

_____

What shape has the correct value?

_____

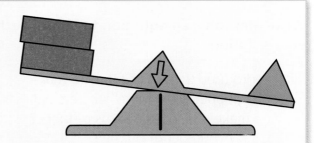

Write a numerical equation to represent the shapes on the scale.

_____

Model 2

How could you use a star and a rectangle to make the scale balance?

_____

_____

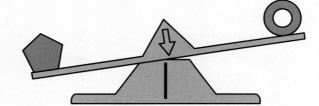

What is the value of each side? _____

Write a numerical equation to represent the shapes on the scale.

_____

 Use the shapes above to draw a balanced scale on a piece of paper. Then write the equation to represent the shapes.

_____

Practice

Draw shapes on each scale to make them balanced. Then write
an equation to match each balanced scale.

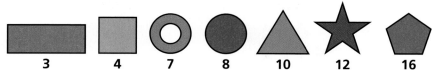

3	4	7	8	10	12	16

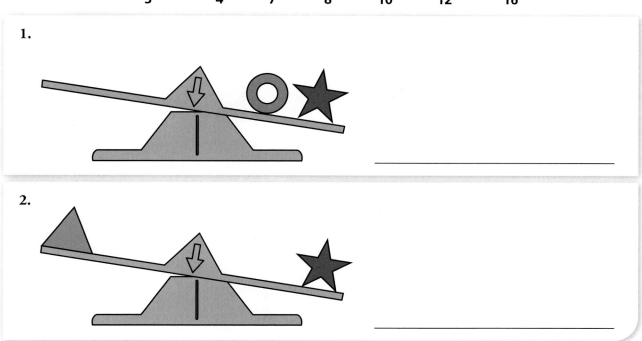

1.

_____

2.

_____

Would a scale be balanced with these shapes? Write *Yes* or *No.*

	Left Side	Right Side	*Yes* or *No*
3.	2 squares and 1 triangle	2 circles	
4.	1 pentagon and 1 donut shape	2 triangles and 3 rectangles	
5.	1 rectangle, 1 square, and 1 circle	1 rectangle, and 1 star	

Draw six of your own shapes. Give each shape a value. Then write
equations with your shapes. Have a partner check your work.

Strength Builder

▶ Function Boxes

Todd made up a game to play with a function box. The game is called *Three and Out.* When a number is put into the box, three things happen. The output from the function box is a new number.

Todd put a 4 into the box. A 20 came out. What three things could have happened in the box?

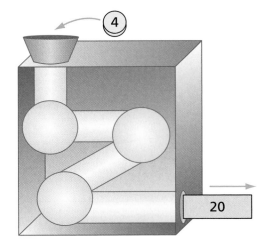

Try these two possible solutions.

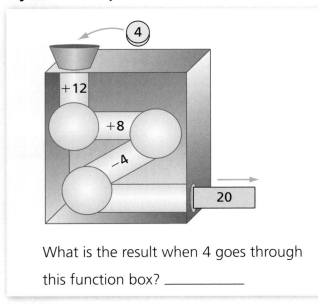

What is the result when 4 goes through

this function box? _____

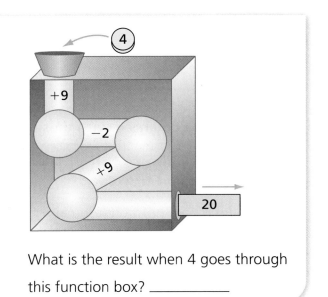

What is the result when 4 goes through

this function box? _____

 What is another set of three addition or subtraction operations that could happen in the function box to turn 4 into 20?

_____

_____

▶ Play the Game

Play *Three and Out* with the numbers below. Write the addition and subtraction that could happen in each box. Hint: There are many ways to find each solution.

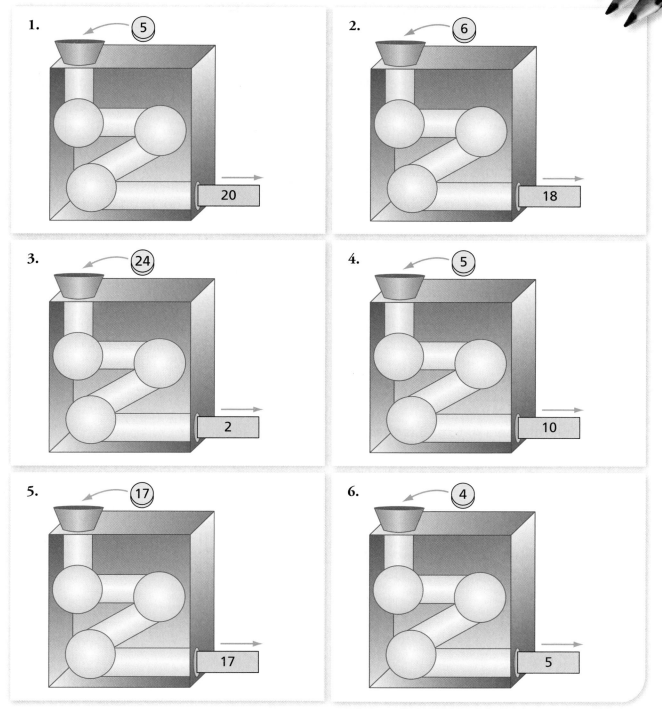

1. 5 → 20

2. 6 → 18

3. 24 → 2

4. 5 → 10

5. 17 → 17

6. 4 → 5

 Use a separate piece of paper. Make your own *Three and Out* function boxes. Write the input and output on the boxes. Then have a partner find a solution for each box.

Write an expression to model each situation. Then find the value of the expression.

1. Manuel had 15 baseball cards. He gave away 4 of the cards.

2. Elizabeth had 8 markers. She bought a box of 12 more markers.

Write an expression with a variable for each situation.

3. There were 6 glasses in the cabinet. Ed placed more glasses in the cabinet.

4. Karla put all the books on the desk into 3 equal piles.

Write and solve an equation for each situation.

5. Tyrell spent $4 for lunch and later bought dinner. He spent $12 in all. How much did he spend for dinner?

6. Wanda had 14 beads. She needed 28 beads to make a necklace. How many more beads did she need?

Which of the numbers 4, 5, or 6 is the solution of each equation?

7. $10 - n = 4$

_____

8. $n + 16 = 20$

_____

Solve the equations. Check your solutions.

9. $n + 33 = 37$

10. $5 \times n = 45$

11. $20 \times 3 = n$

12. $20 + 11 = n$

13. $n - 16 = 20$

14. $42 - n = 32$

Equations and Properties

Do you like

the zoom of a fast roller coaster? There is a roller coaster in Ohio that can travel over 90 miles per hour!

People who build roller coasters use math every day. They must think of speed and safety. They use many formulas to find out how much steel they need. The roller coaster in Ohio has over 19,000 bolts holding it together. How do you think someone figured out the number of bolts they would need?

Commutative Property and Zero Property of Addition

The **Commutative Property of Addition** states that changing the order of addends does *not* change the sum.

Examples: $4 + 9 = 9 + 4$ $7 + x = x + 7$ $a + b = b + a$

Write and solve an addition equation to model each situation.

Tina bought a $4 canvas and an $8 paint set. How much did she spend?	Jerry bought an $8 canvas and a $4 paint set. How much did he spend?
Let n equal the amount of money Tina spent.	Let n equal the amount of money Jerry spent.
$\$4 + \$8 = n$	$\$8 + \$4 = n$
$n =$ _____	$n =$ _____

The addends in the equations are in a different order.
The sums are equal.

$\$4 +$ _____ $=$ _____ $+ \$4$

The **Zero Property of Addition** states that the sum of zero and a number is that same number.

Examples: $0 + 8 = 8$ $5 + 0 = 5$ $y + 0 = y$

James spent 2 hours at soccer practice on Monday. On Tuesday, his team did not practice. How many hours did James spend at soccer practice on Monday and Tuesday?

Let y equal the amount of time James spent at soccer practice on Monday and Tuesday.

_____ $+$ _____ $= y$

$y =$ _____ hours

Explain why the Commutative Property is sometimes referred to as the Order Property.

_____

_____

Practice

Use an addition property to solve and check each equation.

1. $x + 7 = 7$

  $x =$ ___0___

  $0 + 7 = 7$

  $7 = 7$

2. $2 + 5 = s + 2$

  $s =$ _____

3. $11 + 0 = n$

  $n =$ _____

4. $8 + y = 12 + 8$

  $y =$ _____

5. $n + 21 = 21 + 6$

  $n =$ _____

6. $9 + s = 9$

  $s =$ _____

Write an equation with variable n to model each situation. Then solve the equation.

7. Geraldo walked 3 miles and jogged 2 miles from school to the park. On the same route back, he jogged 2 miles and walked the rest of the way. How many miles did Geraldo walk on his way back to school?

  $3 + 2 = 2 + n$

  $n = 3$ miles

8. Winnie used the same number of beads to make two necklaces. She used 9 red beads and 8 green beads on the first necklace. She used 8 green beads on the second necklace. How many red beads did she use on the second necklace?

9. Joe did yard work for 4 hours on Saturday. He didn't do any yard work on Sunday. How many hours of yard work did Joe do on Saturday and Sunday?

10. Tickets to the basketball game were on sale Monday, Tuesday, and Wednesday. Thirty-five tickets were sold in all. On Monday 13 tickets were sold, and on Wednesday 22 tickets were sold. How many tickets were sold on Tuesday?

 Explain how using the Commutative Property of Addition can help you add $4 + 8 + 6$ using mental math.

_____

_____

Associative Property of Addition

The **Associative Property of Addition** states that the grouping of addends does not change the sum of those same addends.

Examples: $(5 + 2) + 3 = 5 + (2 + 3)$ $a + (b + c) = (a + b) + c$

Parentheses are symbols used to group numbers. **Draw parentheses to show the same addends grouped in two different ways. Then find the value of the expressions.**

Model 1

Grouping A	Grouping B
$(4 + 6) + 1 =$	$4 + 6 + 1 =$
_____ + 1	$4 +$ _____
The value of the expression	The value of the expression
is _____.	is _____.

Note that the value of the expression did not change when the grouping of addends changed.

Model 2

Find the value of n. Check the solution.

$(9 + 1) + 8 = 9 + (n + 8)$

Check your solution.

$(9 + 1) + 8 = 9 + (1 + 8)$

_____ $+ 8 = 9 +$ _____

_____ $=$ _____

Explain the reason the Associative Property of Addition is also referred to as the Grouping Property of Addition.

_____

_____

_____

Practice

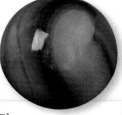

Group the addends in two different ways and find the value of the expressions. Then use the two groupings of addends to write an equation.

1. $3 + 5 + 5 = (3 + 5) + 5$

$\qquad = \underline{\quad 8 \quad} + 5$

The value of the expression is $\underline{\quad 13 \quad}$.

$3 + 5 + 5 = 3 + (5 + 5)$

The value of the expression is $\underline{\qquad}$.

$(3 + 5) + 5 = 3 + (5 + 5)$

2. $20 + 5 + 6 =$

The value of the expression is $\underline{\qquad}$.

$20 + 5 + 6 =$

The value of the expression is $\underline{\qquad}$.

Find the value of _n_. Check your solution.

3. $10 + (7 + 10) = (10 + n) + 10$

$n = 7$

$10 + (7 + 10) = (10 + 7) + 10$

$10 + 17 = 17 + 10$

$27 = 27$

4. $(7 + 8) + 6 = 7 + (8 + n)$

5. $150 + (n + 50) = (150 + 0) + 50$

6. $n + (3 + 25) = (2 + 3) + 25$

Is it easier to find the sum of $(37 + 50) + 50$ or $37 + (50 + 50)$ using mental math? Explain.

_____

_____

Commutative Property of Multiplication

The **Commutative Property of Multiplication** states that changing the order of factors does not change the product of those factors.

Examples: $7 \times 6 = 6 \times 7$ $9 \times n = n \times 9$ $a \times b = b \times a$

Find the value of *n* in each equation.

Model 1

$6 \times n = 48$ $n \times 6 = 48$

$n =$ _____ $n =$ _____

Check: $6 \times$ _____ $= 48$ **Check:** _____ $\times 6 = 48$

_____ $= 48$ _____ $= 48$

The factors in the equations are in a different order, but the products are equal.

$6 \times$ _____ $=$ _____ $\times 6$

Use the Commutative Property to write and solve an equation for the situation.

Model 2

There are 4 rows of drummers marching in the band. Seven drummers are in each row. There is the same number of trumpet players in the band. There are 7 in each row. How many rows of trumpet players are in the band?

Let *n* stand for the number of rows of trumpet players.

Rows of drummers	Drummers in each row	Trumpet players in each row	Rows of trumpet players
_____ ×	_____ =	_____ ×	_____

$n =$ _____. There are _____ rows of trumpet players in the band.

The product of two whole numbers is 12. One of the factors is *n*. Use the Commutative Property to write two equations using those facts.

_____

_____

Practice

**Find the value of *n* in each equation. Check your solutions.
Then use the factors to write a new equation that shows the
Commutative Property of Multiplication.**

1. $4 \times n = 36$ $n \times 4 = 36$
       $n = 9$ $n = 9$
     $4 \times 9 = 36$ $9 \times 4 = 36$
         $4 \times 9 = 9 \times 4$

2. $n \times 11 = 33$ $11 \times n = 33$

3. $n \times 10 = 100$ $10 \times n = 100$

4. $12 \times n = 60$ $n \times 12 = 60$

Find the value of *y*.

5. $7 \times 8 = y \times 7$
     $y = 8$

6. $y \times 6 = 6 \times 5$

7. $14 \times y = 2 \times 14$

8. $25 \times 25 = 25 \times y$

Write and solve an equation to model each situation.

9. There are 6 desks in each of 4 rows in Rita's class. Juan's class has the same number of desks. They are in 6 equal rows. How many desks are in each row in Juan's class?

Let n = the number of desks in each row in Juan's class.

10. Kevin put 5 roses in each of 7 vases. Glenda placed the same number of roses in 7 vases. How many roses did Glenda put in each vase?

Let n = the number of roses in each of Glenda's vases.

 How is the Commutative Property of Addition similar to the Commutative Property of Multiplication?

_____

_____

Associative Property of Multiplication

The **Associative Property of Multiplication** states that changing the grouping of factors does not change the product of those factors.

Examples: $(6 \times 5) \times 2 = 6 \times (5 \times 2)$ $a \times (b \times c) = (a \times b) \times c$

Parentheses are symbols used to group numbers. **Draw parentheses to show the same factors grouped in two different ways. Then find the value of the expressions.**

 Model 1

Grouping A	Grouping B
$(3 \times 5) \times 2 =$	$3 \times 5 \times 2 =$
_____ $\times 2$	$3 \times$ _____
The value of the expression	The value of the expression
is _____.	is _____.

Note that the value of the expression did not change when the grouping of addends changed.

Use the Associative Property to find the value of *n*.

Model 2

$3 \times (4 \times 2) = (3 \times n) \times 2$

$n =$ _____

Check your solution. $3 \times (4 \times 2) = (3 \times 4) \times 2$

$3 \times$ _____ $=$ _____ $\times 2$

_____ $=$ _____

Explain how the Associative Property of Multiplication can help you compute 3 × 8 × 5 using mental math.

_____

_____

Practice

Group the addends two different ways and find the value of the expressions. Then use the two groupings of factors to write an equation.

1. $5 \times 3 \times 2 = (5 \times 3) \times 2$ $5 \times 3 \times 2 = 5 \times (3 \times 2)$

$= \underline{\quad 15 \quad} \times 2$ $= 5 \times \underline{\hspace{2cm}}$

The value of the expression is $\underline{\quad 30 \quad}$.   The value of the expression is $\underline{\hspace{2cm}}$.

$(5 \times 3) \times 2 = 5 \times (3 \times 2)$

2. $10 \times 5 \times 2 =$ $10 \times 5 \times 2 =$

$\underline{\hspace{2cm}}$ $\underline{\hspace{2cm}}$

The value of the expression is $\underline{\hspace{2cm}}$.   The value of the expression is $\underline{\hspace{2cm}}$.

Use the Associative Property to find the value of n. Check your solution.

3. $10 \times (2 \times 7) = (10 \times n) \times 7$

$n = 2$

$10 \times (2 \times 7) = (10 \times 2) \times 7$

$10 \times 14 = 20 \times 7$

$140 = 140$

4. $(2 \times 6) \times 2 = 2 \times (6 \times n)$

5. $(3 \times n) \times 9 = 3 \times (3 \times 9)$

6. $n \times (5 \times 2) = (8 \times 5) \times 2$

 In your own words explain how the Commutative Property of Multiplication and the Associative Property of Multiplication are different.

$\underline{\hspace{14cm}}$

$\underline{\hspace{14cm}}$

$\underline{\hspace{14cm}}$

Zero Property and Identity Property of Multiplication

The **Zero Property of Multiplication** states that when zero is a factor, the product is zero. If the product in multiplication is zero, then at least one of the factors is zero.

Examples: $0 \times 12 = 0$ $n \times 6 = 0$ $4 \times n \times 7 = 0$
                                        $n = 0$ $n = 0$

Use the Zero Property of Multiplication to find each missing value.

Model
If $17 \times n = 0$, then $n =$ _____.   If $5 \times 0 = n$, then $n =$ _____.
The missing factor must be zero   The product must be zero because
because the product is zero.   one of the factors is zero.

The **Identity Property of Multiplication** states that when 1 is multiplied by a number, the product is that number. This property is also referred to as the **Property of One**.

Examples: $1 \times 33 = 33$ $n \times 6 = 6$ $3 \times n \times 9 = 27$
                                        $n = 1$ $n = 1$

Use the Identity Property of Multiplication to find each missing value.

Model 2

If $8 \times n = 8$, then $n =$ _____.   If $25 \times 1 = n$, then $n =$ _____.
The missing factor must be 1   The product is 25 because a
because the product is the same   number times 1 is itself.
as the number multiplied.

Write an equation with variable n to model the situation. Then solve the equation.

Model 3
Darryl bought 12 marbles for $1 each. What was the total cost of the marbles? Let n represent the total cost of the marbles.

$12 \times \$1 = n$ $n =$ _____

Explain how the problem $1 \times 0 = 0$ shows both the Zero Property and the Identity Property of Multiplication.

_____

_____

Practice

Find the value of *n*. Check your solution.

1. $n \times 12 = 0$

$n = \underline{\quad 0 \quad}$

$0 \times 12 = 0$

$0 = 0$

2. $81 \times n = 81$

$n = \underline{\quad\quad}$

3. $1 \times 0 = n$

$n = \underline{\quad\quad}$

4. $34 \times n = 1 \times 34$

$n = \underline{\quad\quad}$

5. $n \times 17 \times 2 = 0$

$n = \underline{\quad\quad}$

6. $1 \times 77 \times n = 77$

$n = \underline{\quad\quad}$

7. $24 \times n \times 2 = 48$

$n = \underline{\quad\quad}$

8. $n \times 15 = 15 \times 0$

$n = \underline{\quad\quad}$

9. $18 \times 3 \times 0 = n$

$n = \underline{\quad\quad}$

Write an equation with variable *n* to model each situation. Then solve the equation.

10. Elisa gave one party favor to each of the 12 guests at her party. How many party favors did Elisa give to her guests?

Let *n* = the number of party favors Elisa gave to her guests.

$1 \times 12 = n$

$n = 12$ party favors

11. There are 6 glasses for fruit juice. Each glass is empty. How many ounces of fruit juice are in the glasses?

Let *n* = the number of ounces of fruit juice.

What property could help you solve this equation quickly? Explain. $36 \times 772 \times 88 \times 994 \times 0 \times 456 = y$

_____

_____

_____

_____

_____

Lesson 6
Distributive Property

To **distribute** means to give something to each member of a group. The **Distributive Property** states that when a number is being multiplied by a sum or difference, that number is distributed to each addend in the sum.

Simplify the expression $5 \times (3 + 4)$.

Examples: Use the Distributive Property.

$$5 \times (3 + 4) =$$
$$(5 \times 3) + (5 \times 4) =$$
$$15 + 20 =$$
$$\underline{\hspace{2cm}}$$

Simplify each expression using the distributive property.

Model 1

$2 \times (8 + 5) =$	$6 \times (3 - 2) =$
$(2 \times 8) + (2 \times \underline{\hspace{1cm}}) =$	$(6 \times 3) - (6 \times \underline{\hspace{1cm}}) =$
$\underline{\hspace{1cm}} + \underline{\hspace{1cm}} =$	$\underline{\hspace{1cm}} - \underline{\hspace{1cm}} =$
$\underline{\hspace{1cm}}$	$\underline{\hspace{1cm}}$

Use the Distributive Property to simplify the expression.

Model 2

$12 \times (n + 2)$ Let $n = 3$.

$$12 \times (n + 2) =$$
$$12 \times (3 + 2) =$$
$$(12 \times \underline{\hspace{1cm}}) + (12 \times \underline{\hspace{1cm}}) =$$
$$\underline{\hspace{1cm}} + \underline{\hspace{1cm}} =$$
$$\underline{\hspace{1cm}}$$

 Create a drawing using geometric figures to demonstrate the Distributive Property.

Practice

Simplify each expression using the Distributive Property.

1. $3 \times (7 + 2)$ $(3 \times 7) + (3 \times 2) =$ $21 + 6 =$ 27	**2.** $6 \times (2 + 1)$	**3.** $11 \times (3 - 1)$
4. $8 \times (0 + 9)$	**5.** $4 \times (4 + 3)$	**6.** $20 \times (7 - 3)$
7. $5 \times (10 - 3)$	**8.** $7 \times (2 + 5)$	**9.** $6 \times (5 + 4)$

Use the Distributive Property to simplify each expression.

10. $4 \times (n + 4)$ if $n = 5$. $4 \times (5 + 4) =$ $(4 \times 5) + (4 \times 4) =$ $20 + 16 =$ 36	**11.** $25 \times (n + 2)$ if $n = 1$.	**12.** $n \times (6 - 4)$ if $n = 9$.
13. $n \times (0 + 5)$ if $n = 12$.	**14.** $2 \times (10 - n)$ if $n = 8$.	**15.** $8 \times (5 + n)$ if $n = 3$.

 A classmate wrote $5 \times (9 + 3) = 5 \times 9 + 3$. Explain the error and use the Distributive Property to write the statement correctly.

_____

_____

Lesson 7 — Order of Operations

In a problem with more than one operation, there is a specific order to follow. This order is called the **Order of Operations**.

Order of Operations

1. Perform all operations inside parentheses.
2. Multiply or divide from left to right.
3. Add or subtract from left to right.

Examples:

$2 + 10 \div 5$	$3 \times (12 - 8)$
$2 + \mathbf{10 \div 5}$ *Divide*	$3 \times \mathbf{(12 - 8)}$ *Parentheses*
$\mathbf{2 + 2}$ *Add*	$\mathbf{3 \times 4}$ *Multiply*
4	12

Simplify each expression using the Order of Operations.

Model 1

$3 + 4 \times 9$ Which step is first?	$8 \times (3 + 6)$ Which step is first?
_____	_____
$3 + 4 \times 9$ Multiply	$8 \times (3 + 6)$ Parentheses
$3 + $ _____ Add	$8 \times $ _____ Multiply
_____	_____

Model 2

$2 \times (9 - 4) + 3$	$8 \times 3 + 6 - 4$
Which step is first?	Which step is first?
_____	_____
$2 \times (9 - 4) + 3$ Parentheses	$8 \times 3 + 6 - 4$ Multiply
$2 \times $ _____ $+ 3$ Multiply	$24 + 6 - 4$ Add
$10 + 3$ Add	_____ $- 4$ Subtract
_____	_____

 Explain how to simplify $12 \div 2 \times 3$ using the Order of Operations.

_____

_____

_____

Practice

Write which step is first. Then find the value of each expression.

1. $20 - 9 \times 2$

multiplication

$20 - 18 = 2$

2. $4 + 35 \div 7$

3. $(6 + 6) \div 3$

4. $5 \times (20 - 10)$

5. $(56 \div 8) \times 7$

6. $35 - 15 + 8$

7. $4 + (6 \times 6)$

8. $(5 + 5) \times 2$

9. $4 \times 8 - 30$

Simplify each expression. Show your work.

10. $7 \times (6 + 2) + 1$

$7 \times 8 + 1$

$56 + 1$

57

11. $15 \div 5 - 1 \times 2$

12. $64 \div (8 \times 1) - 3$

13. $42 \div (4 + 3) \times 2$

14. $(20 - 2) \div 9$

15. $8 + 5 \times 4 - 2$

16. $10 - 2 \times 3 \div 3$

17. $(2 + 3) \times 4 - 4$

18. $2 + (18 - 9) \div 3$

 Simplify $5 \times (3 + 6)$ using the Distributive Property and the Order of Operations. What do you notice about the value of the expressions using both methods?

_____

_____

Strength Builder

▶ Expression Game

The game *Make the Most of It* can be played by 2 or more players.
Players try to make the greatest possible value using the numbers rolled
and the available symbols.

Materials

- 4 index cards
- Scorecard for each player
- 1 number cube
- Pencil for each player

Getting Ready

1. Write an addition sign on one index card. (+) Use three other
   index cards to write symbols for subtraction, multiplication,
   and division. (−, ×, ÷) Turn the cards face down, mix them
   up, and stack them between the players.

2. To make a number cube, copy or trace the figure at right.
   Write a number from 1 to 6 in each square. Cut out the
   figure, fold it, and tape the tabs to make a cube.

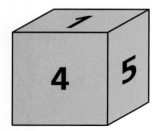

3. Use the scorecard on the bottom of page 35 or copy it
   onto another piece of paper. Make sure every player has
   a scorecard.

▶ Playing the Game

1. Have each player roll the number cube. The player who rolls the greatest number will be first.

2. The first player rolls the number cube three times. All players will write the numbers on their scorecards in the first three spaces beside *Round 1.*

3. Then the first player draws two symbols cards. All players will write the symbols on their scorecards.

4. Each player will write an expression that contains the three numbers and two symbols for this round.

5. Finally, all players will find the value of the expression each of them wrote. Calculators can be used to check answers. Each player will record his or her answer in the *Score* column of the scorecard.

6. At the end of five rounds of play, each player will total his or her own score. The greatest total score wins.

Game Hint

Suppose the numbers rolled are 3, 4, and 5, and the symbols drawn are $\times$ and $+$. Which expression would make the highest score?

$3 \times 4 + 5 = $ _____ $\qquad$ $4 \times 5 + 3 = $ _____

The object of the game is to write expressions that will give you the greatest score.

Make the Most of It Scorecard						
Round	**Numbers Rolled**			**Symbols Drawn**		**Expression**	**Score**
1							
2							
3							
4							
5							
						TOTAL:	

Variation

Another game can be played using the same materials, *Make the Least of It.* Use the same scorecard and write expressions that will result in the lowest score possible. At the end of five rounds, the player with the lowest total score will win.

Solve each equation for _y_. Name the addition property you used.

1. $y + 9 = 9$

2. $3 + 7 = y + 3$

3. $(6 + 7) + 2 = y + (7 + 2)$

4. $0 + y = 13$

Solve each equation for _n_. Name the multiplication property you used.

5. $4 \times (10 \times 3) = (4 \times n) \times 3$

6. $5 \times (2 + 8) = (5 \times 2) + (n \times 8)$

7. $8 \times n \times 6 = 0$

8. $17 \times 2 \times n = 34$

Simplify each expression.

9. $18 - 4 \times 3$

10. $7 + 9 \div 3$

11. $6 \times 3 \div 9 + 4$

12. $36 \div (4 + 5)$

13. $(15 - 5) \div 2$

14. $(4 + 3) \times 6 + 1$

Write and solve an equation to model each situation.

15. There are 4 rows of corn in Chen's garden. Each row has 3 plants. Pat's garden has the same amount of corn. There are 3 rows of plants. How many plants are in each row of Pat's garden?

Let n = the number of plants in each row of Pat's garden.

16. Erin practiced the piano for 1 hour each day. She practiced a total of 7 hours. How many days did Erin practice?

Let n = the number of days Erin practiced the piano.

Representations and Functions

How many pictures can you take in a second?

A motion picture camera takes 24 pictures per second! That's 1,440 pictures per minute!

When you see a two-hour movie, it is made of over 172,000 pictures, called frames. In a movie theater, a projector uses two flashes of light for each frame. If 1 frame equals 2 flashes of light, how many flashes of light are there in a minute?

1 Data Tables

Some students voted for their favorite color and recorded their data in a table. A **table** shows the data in an organized way.

Examples:

Favorite Color			
Color	**Red**	**Blue**	**Green**
Number of Votes	12	15	8

Record all the combinations of coins that add up to 15¢.

Model 1 ▶ What categories are used to classify the coins?

_____

How many different combinations of coins add up to 15¢?

_____

Coins Equal to 15¢		
Dimes	**Nickels**	**Pennies**
1	1	

Model 2 ▶ Students voted for their favorite activity from a list of volleyball, swimming, and biking. **Make a table from their tally chart.**

Volleyball	||
Swimming	ЖŦ I
Biking	||||

Favorite Activity			
Activity			
Number of Votes			

Explain how the table makes the data easy to read and understand.

_____

_____

Practice

Use the table to record the combinations of bills that add up to $18.

Bills Equal to $18		
$10 bills	$5 bills	$1 bills
1. 1	1	3
2.		
3.		
4.		
5.		
6.		

7. What categories are used to classify the bills? _____

8. What combination of six bills is equal to $18? _____

9. Which combination uses the greatest number of bills? _____

10. Three groups of students in a class voted for a favorite type of movie. **Make a table that combines the data from their tally charts.**

Group 1	Group 2	Group 3
Action ‖	Action ‖‖	Action ‖‖
Comedy		Comedy	Comedy	
Cartoon ‖‖	Cartoon ‖‖	Cartoon
Scary	Scary	Scary ‖‖‖

Favorite Type of Movie				
Type of Movie				
Number of Votes				

 Conduct a survey about a topic that interests you. Make a table on a separate piece of paper to display the results of the survey.

Line Graphs

There are many different models you can use to represent data. A **line graph** is the best model to use to show changes over time.

Model 1

James recorded the average weekly temperatures at his home in Houston one December.

Use the line graph to answer the questions.

In which weeks did the temperature decrease from the previous week?

_____

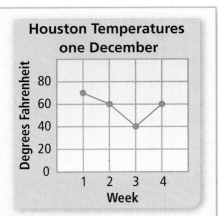

Model 2

Brad recorded the profit he made selling apples for four days.

Profit Selling Apples				
Day	Monday	Tuesday	Wednesday	Thursday
Amount	$75	$50	$75	$100

Make a line graph. Plot the data point for each day given in the table.

The first point is (Mon, $75).

The second point is _____.

The third point is _____.

The last point is _____.

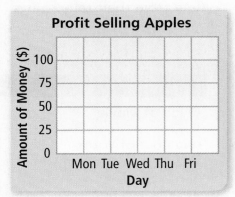

Connect the points to see the change in the profit Brad made.

Look at the line graph for Brad's profits. How does it show which days the amount of money increased or decreased?

_____

_____

Practice

The table shows the height of a sunflower on the first day of each month. Answer the questions using the line graph.

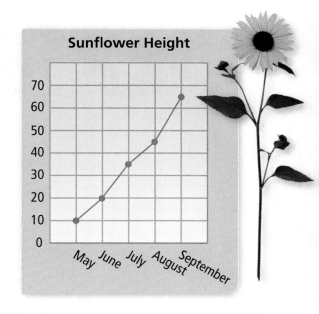

Sunflower Height	
Month	Height (in.)
May 1	10
June 1	20
July 1	35
August 1	45
September 1	65

1. Did the height of the plant increase or decrease from May through September?

_____increase_____

2. In which month did the sunflower grow the most?

_____

3. During which two months did the sunflower grow the same amount?

_____

4. What was the height of the sunflower July 1?

_____

Make a line graph to show the number of hours Kristy spent at her computer last week.

5.

Computer Time	
Day	Number of Hours
Mon	1
Tues	1.5
Wed	3
Thu	2.5
Fri	2.5

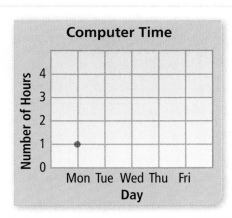

 How do you know that there is not a change in value between two points when looking at a line graph?

_____

_____

3 Function Tables

A rule that shows the relationship between two variables such as x and y is called a **function**. For each value of x you input, there is only one output for y. You can represent this relationship in a **function table**.

If the rule is add two, the function table might look like this.

Example:

Input x	Function $y = x + 2$	Output y
1	$y = 1 + 2$	3
2	$y = 2 + 2$	4

The function table below shows the relationship between hours worked (h) and money earned (m).

Model 1 ▶ **Complete the table.**

Input h (hours worked)	Function $m = h \times \$6$	Output m (money earned)
1	$m = 1 \times \$6$	$6
2	$m = \underline{\hspace{1cm}} \times \$6$	
3	$m = \underline{\hspace{1cm}} \times \$6$	

Find the relationship between hours (h) and distance (d).

Model 2 ▶ What happens to h to produce d?

_____

Input h (hours driven)	Output d (distance driven)
1	50 miles
2	100 miles
3	150 miles

What is the function that shows the relationship between h and d?

output	=	input	×	50
_____	=	_____	×	50

The function $y = x \div 5$ is given. What is the rule? Which variable would represent the input in the table?

_____

Practice

The price of oranges is $2 a pound. The function table below shows the relationship between pounds of oranges bought (*a*) and money paid (*m*). Complete the table.

Input a (pounds of oranges)	Function m = a × $2	Output m (money paid)
1	m = ___1___ × $2	$2
2	m = _____ × $2	
3	m = _____ × $2	
4	m = _____ × $2	
5	m = _____ × $2	

6. What is the value of *m* when *a* is 3? _____

7. What is the value of *a* when *m* is $10? _____

8. What will the value of *m* be when *a* is 9? _____

9. The function table below shows the relationship between the number of television commercials (*c*) and seconds (*s*).

Input c (commercials)	Output s (seconds)
1	30
2	60
3	90

What is the function?

_____

10. The function table below shows the relationship between the number of hours (*h*) and the distance (*d*) Jan biked.

Input h (hours)	Output d (distance)
1	6 miles
2	12 miles
3	18 miles
4	24 miles

What is the function?

_____

Lamar charges $10 an hour to baby-sit. He works 1 to 6 hours at a time. Make a function table on a separate piece of paper. Show the relationship between the number of hours he baby-sits (*h*) and the amount of money he makes (*m*).

Graphs and Function Tables

You can show the graph of a function on a **coordinate grid**. A coordinate grid contains an **x-axis** (horizontal axis) and a **y-axis** (vertical axis).

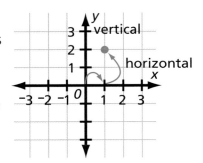

An **ordered pair** (x, y) gives the location of a point on the grid. The first number is the x-value and tells you how far to move on the x-axis from zero. The second number is the y-value and tells you how far to move on the y-axis.

The ordered pair (1, 2) represents moving right 1 from 0 on the x-axis and up 2 on the y-axis.

Write each ordered pair.

Model 1

Input x	Function y = x + 3	Output y	Ordered Pair (x, y)
0	y = 0 + 3	3	(0, 3)
2	y = _____ + 3	5	
4	y = _____ + 3		(4, 7)
6	y = _____ + 3		

Graph and label the points of the function $y = x + 1$.

Model 2

Input x	Function y = x + 1	Output y	Ordered Pair (x, y)
0	y = 0 + 1	1	(0, 1)
1	y = 1 + 1	2	(1, 2)
3	y = 3 + 1	4	(3, 4)
4	y = 4 + 1	5	(4, 5)

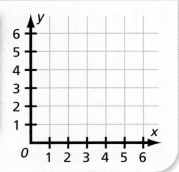

Explain how to graph the points (0, 1) and (2, 0) on the coordinate grid.

_____

_____

Write each ordered pair.

1.

Input x	Function y = 5 + x	Output y	Ordered Pair (x, y)
1	y = 5 + ___1___	6	(1, 6)
3	y = 5 + _____		
5	y = 5 + _____		
7	y = 5 + _____		

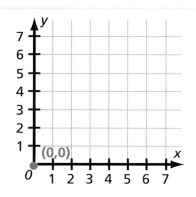

2.

Input x	Function y = 4x	Output y	Ordered Pair (x, y)
0	y = 4 × _____		
2	y = 4 × _____		
4	y = 4 × _____		
6	y = 4 × _____		

Graph and label the points of the function y = 2x.

3.

Input x	Function y = 2x	Output y	Ordered Pair (x, y)
0	y = 2 × 0	1	(0, 0)
1	y = 2 × 1	2	(1, 2)
2	y = 2 × 2	4	(2, 4)
3	y = 2 × 3	6	(3, 6)

 The function is *y* = 2*x*. If *x* = 5, what ordered pair would you graph on the coordinate grid? Explain how you would graph that point on a coordinate grid.

_____

_____

_____

Lesson 5

Tables of Solution Sets

The equation $y = x - 8$ has two variables, x and y. Any ordered pair (x, y) that makes the equation true is a **solution** of the equation. Because there is more than one solution, all of the possible solutions form a **solution set**. You can organize the solutions in a table.

Example: $y = x - 8$

Input x	Equation y = x − 8	Output y	Ordered Pair (x, y)
15	y = 15 − 8	7	(15, 7)
39	y = 39 − 8	31	(39, 31)
80	y = 80 − 8	72	(80, 72)

Find solutions of the equation $y = 3 + x - 4$.

Model 1

Input x	Equation y = 3 + x − 4	Output y	Ordered Pair (x, y)
6	y = 3 + 6 − 4	5	(6, 5)
11	y =		
20	y =		
29	y =		

To see if an ordered pair is a solution of an equation, substitute the values into the equation and make sure both sides of the equation are equal.

Model 2 Is (3, 9) a solution of the equation $y = 3x$? **Substitute 3 for x, and 9 for y in the equation.**

$y = 3x$

$9 = 3 \times 3$

$9 = 9$ The pair (3, 9) is a solution of $y = 3x$.

 Is (5, 8) a solution to the equation $y = 3x$? Explain.

_____

Practice

Find four solutions of each equation. Write the solutions as ordered pairs.

	Input x	Equation y = x ÷ 4	Output y	Ordered Pair (x, y)
1.	4	y = <u>4 ÷ 4</u>	1	(4, 1)
2.	8	y = _____		
3.	16	y = _____		
4.	40	y = _____		

	Input x	Equation y = 2x + 1	Output y	Ordered Pair (x, y)
5.	1	y = <u>2 × 1 + 1</u>	3	(1, 3)
6.	3	y = _____		
7.	5	y = _____		
8.	7	y = _____		

Determine if the ordered pair is a solution of the given equation. Show your work.

9. $(2, 9)$ $y = x + 7$

$y = x + 7$

$9 = 2 + 7$

$9 = 9$

Yes

10. $(12, 4)$ $y = 3x$

11. $(5, 11)$ $y = 2x + 3$

12. $(30, 5)$ $y = x \div 6$

13. $(1, 1)$ $y = x$

14. $(14, 5)$ $y = x - 8$

 Write an equation with variables *x* and *y*. Find four solutions of the equation.

_____

_____

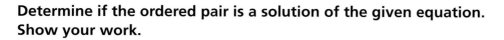

Graphs of Solution Sets

The solutions for an equation with
two variables can be represented
on a graph. This graph shows the
solutions for $y = x + 2$.

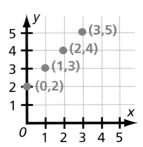

Example:

Graph four solutions of the equation $y = 12 - x$.

Model 1 ▶ Find the solutions.
Then write the
solutions as
ordered pairs.
Use 2, 5, 8, and
10 for x-values.

x	$y = 12 - x$	y	(x, y)
2	$y =$ _12 − 2_	10	(2, 10)
5	$y =$ _____		
8	$y =$ _____		
10	$y =$ _____		

Graph the four
solutions you
found on the
coordinate plane
at right. Label
the points on the
graph. Connect
the points with
a line.

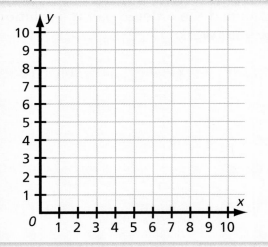

Use the graph of $y = 12 - x$ to find other possible solutions.

Model 2 ▶ Other solutions to the equation will lie on the line of the graph.

When $x = 7$, $y =$ _____. _____ is also a solution of the equation.

  In the equation $y = 2x$, what is y when x is 3? Explain.

_____

_____

Practice

Complete the tables and graph the solutions to the equations.

	x	y = x + 5 − 1	y	(x, y)
1.	1	y = __1 + 5 − 1__	5	(1, 5)
2.	3	y = _____		
3.	6	y = _____		
4.	7	y = _____		
5.	9	y = _____		

6.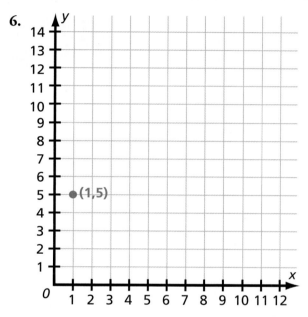

Use the graph of y = x + 5 − 1 to find other solutions of the equation.

7. When x = 2, y = _____

8. When x = 8, y = _____

9. When x = 4, y = _____

10. When x = 0, y = _____

 Explain how you can use the graph of a solution set to find the x-coordinate of a solution when the y-coordinate of the solution is given.

_____

_____

_____

Lesson 7

Linear Equations

When the solutions of an equation lie on a line in a graph, the equation is a **linear equation**.

Complete the table and then graph the solutions.

Model 1 ▶ Carol is buying tennis balls. Each can she buys contains three tennis balls. Let x equal the number of tennis balls she buys and let y equal the number of cans she buys. Try 3, 6, 9, 12, and 15 for x-values.

x (number of tennis balls)	Linear equation $y = x \div 3$	y (number of cans)	Ordered Pair (x, y)
3	$y = 3 \div 3$	1	(3, 1)
6	$y = 6 \div 3$		
9	$y = $ _____		
12	$y = $ _____		
15	$y = $ _____		

Graph the solutions. Then connect the points.

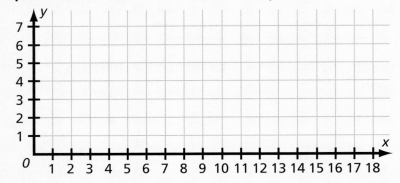

The equation $y = x \div 3$ is a _____ equation because all the solutions lie on the same line.

 Find another solution to the linear equation $y = x \div 3$. Then explain how you found your solution.

_____

_____

Practice

Write an equation to represent each situation. Then graph the solutions.

All merchandise at Sports Plus is $2 off the original price. Let *x* equal the original price and *y* equal the sale price.

	x (original price)	Linear equation $y = x - \$2$	*y* (sale price)	Ordered Pair (*x, y*)
1.	$4	$y =$ $4 − $2	$2	($4, $2)
2.	$6	$y =$ _____		
3.	$8	$y =$ _____		
4.	$9	$y =$ _____		
5.	$10	$y =$ _____		
6.	$11	$y =$ _____		
7.	$12	$y =$ _____		

8.–14.

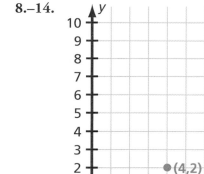

 How do you know by graphing if one of the points you found is not a solution to the linear equation?

_____

_____

Strength Builder

▶ Grid Game

Two or more players can play the game *Greatest Distance*. Players try to plot points that are the greatest distance apart on a grid.

Materials

- 2 number cubes
- Scorecard for each player
- Scraps of colored paper
- Grid paper for each player
- Pencil for each player
- Scissors or hole-punch

Getting Ready

1. Use two number cubes or make your own. Directions for making number cubes are on page 34.

2. Cut or use a hole-punch to make dots of colored paper. These will be used as markers to show points on the game grid. Each player will need two markers.

3. Use the scorecard on the bottom of page 53 or copy it onto another piece of paper. Make sure every player has a scorecard.

4. Make a game board with grid paper for each player. Use the grid below or make your own. Label the *x*-axis and the *y*-axis. Label numbers 0 through 6 on both the *x*-axis and the *y*-axis.

▶ Playing the Game

1. Have one of the players roll the number cubes. You may have the same player roll the cubes each time or take turns.

2. When the cubes are rolled, each player will write a coordinate pair on his or her scorecard. For example, if the numbers 2 and 5 are rolled, players may write (2, 5) or (5, 2) as starting points.

3. Each player marks a starting point on his or her game grid.

4. Roll the cubes again. Write another coordinate pair on your scorecard. Mark the point on your grid. Try to choose the point farthest away from your first point.

   For example, suppose your first point is (2, 5) and the next numbers rolled are 1 and 6. Which point is the greatest distance, (1, 6) or (6, 1)?

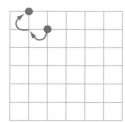 From (2, 5) to (1, 6) is one step across and one step up. 1 + 1 = 2.

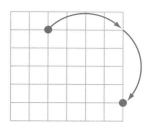 From (2, 5) to (6, 1) is four steps across and four steps down. 4 + 4 = 8. This player would have the higher score for the round.

5. Play the game for five rounds. Record your score for each round. Then add to find your total score. The player with the greatest score wins the game.

Greatest Distance Scorecard				
Round	Point 1	Point 2	Add the Distance	Score
Example	(2, 5)	(6, 1)	4 + 4	8
1				
2				
3				
4				
5				
			FINAL SCORE:	

Make a table to solve. Then show the data on a line graph.

1. Ming ran 2 miles every day. How many miles in all had she run at the end of day 1, day 2, day 3, and day 4?

Ming's Miles	
Day	Miles in All

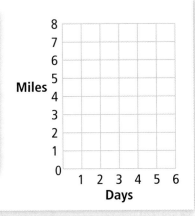

2. Mike is selling peanuts at $3 each pound. How much is it for 0, 1, 2, and 3 pounds of peanuts?

Cost of Peanuts	
Pounds	Cost

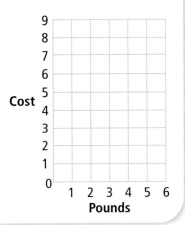

Complete the table for the equation $y = 2 + x$. Graph the solutions and label the points. Then draw the line of the equation.

	Input x	Linear equation $y = 2 + x$	Output y	Ordered Pair (x, y)
3.	0	$y =$ _____		
4.	1	$y =$ _____		
5.	2	$y =$ _____		
6.	3	$y =$ _____		
7.	4	$y =$ _____		

8.

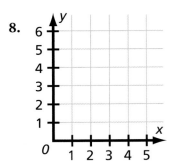

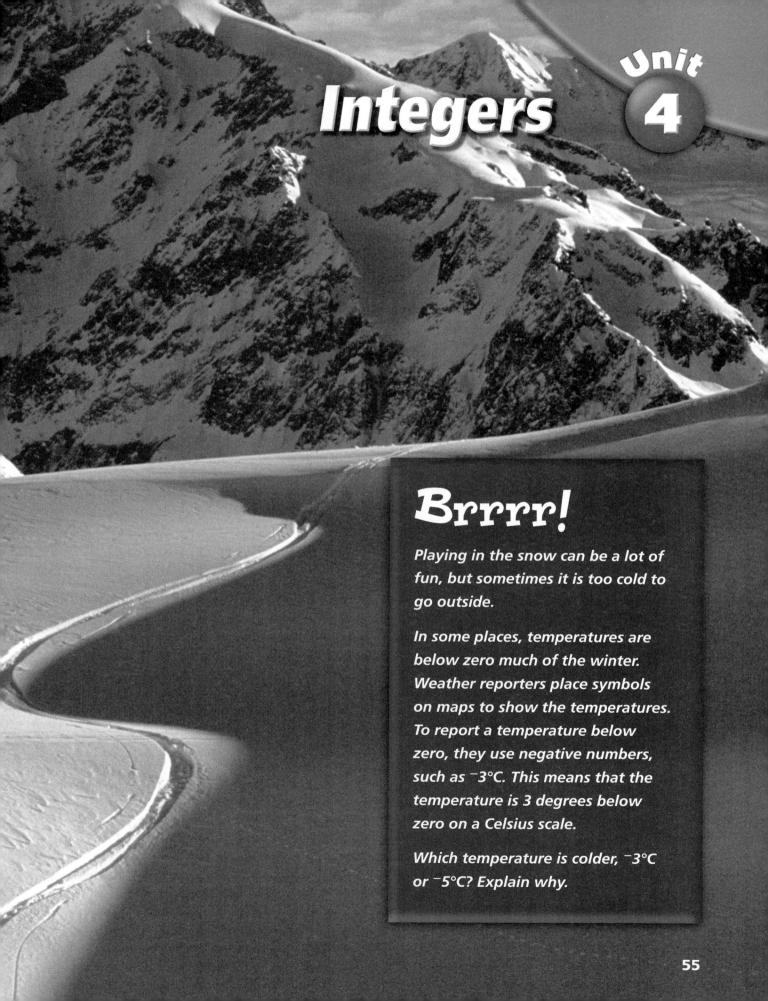

Integers

Brrrr!

Playing in the snow can be a lot of fun, but sometimes it is too cold to go outside.

In some places, temperatures are below zero much of the winter. Weather reporters place symbols on maps to show the temperatures. To report a temperature below zero, they use negative numbers, such as $^-3°C$. This means that the temperature is 3 degrees below zero on a Celsius scale.

Which temperature is colder, $^-3°C$ or $^-5°C$? Explain why.

Lesson 1

Integers and the Number Line

The set of **integers** includes both whole numbers and their opposites. **Opposite numbers** are the same distance from zero on the number line.

Negative numbers are less than zero. **Examples:** ⁻3, ⁻1, ⁻5

Positive numbers are greater than zero. **Examples:** 7 or ⁺7, 2 or ⁺2

Zero is an integer that is neither positive nor negative.

The integers ⁻3 and 3 are opposites.

Graph integers on a number line. Point *F* on the number line shows the number of yards lost on the Wildcats' first play in a football game.

Model 1

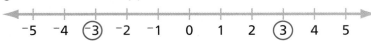

Is the integer at point *F* less than or greater than 0? _____

Is the integer at point *F* a positive or negative integer? _____

How many units from 0 is point *F*? _____

What integer does point *F* show? _____

On another play in the game, the Wildcats gained 4 yards.

Model 2

Is 4 less than or greater than 0? _____

Is 4 a positive or negative integer? _____

Graph 4 on the number line and label it *S*.

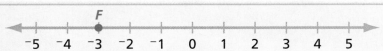

 Explain why $⁻4\frac{1}{2}$ **is not an integer.**

_____

_____

_____

Practice

Write the integer that each point represents.

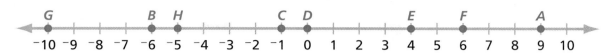

1. point *A* 9	**2.** point *B*	**3.** point *C*	**4.** point *D*
5. point *E*	**6.** point *F*	**7.** point *G*	**8.** point *H*

Graph each integer on the number line below.

9. point *J*, ⁻2	**10.** point *K*, 1	**11.** point *M*, ⁻8	**12.** point *N*, ⁻3
13. point *P*, 7	**14.** point *R*, ⁻5	**15.** point *S*, 3	**16.** point *T*, ⁻7

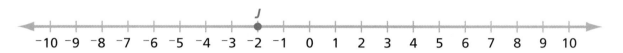

Is each pair of integers opposites? Write *Yes* or *No*.

17. 3 and ⁻3 **Yes**	**18.** 8 and ⁻9	**19.** 7 and ⁻7
20. 0 and 2	**21.** ⁻1 and 1	**22.** ⁻5 and ⁻5

What is the opposite of each integer?

23. 4 ⁻4	**24.** ⁻6	**25.** 10
26. ⁻2	**27.** ⁻9	**28.** 8

 Why can two positive integers never be opposites?

_____

_____

_____

_____

_____

_____

2 Absolute Value

The **absolute value** of an integer is its distance from 0. The symbol for absolute value is two bars placed around the number.

Examples: $|^-2| = 2$ Read *the absolute value of negative two equals two.*

$|3| = 3$ Read *the absolute value of three equals three.*

Find the absolute value of 5.

Model 1

number line: $^-5$ $^-4$ $^-3$ $^-2$ $^-1$ 0 1 2 3 4 5

Mark a point at 5 on the number line.

How many units from 0 is 5? _____

What is the absolute value of 5? _____ $|5| =$ _____

Find the absolute value of $^-5$.

Model 2

number line: $^-5$ $^-4$ $^-3$ $^-2$ $^-1$ 0 1 2 3 4 5

Mark a point at $^-5$ on the number line.

How many units from 0 is $^-5$? _____

What is the absolute value of $^-5$? _____ $|^-5| =$ _____

Find the value of each side of the equation.

Model 3

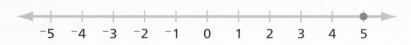

$|7| = |^-7|$ $|^-1| =$ _____

_____ = _____   _____ = _____

Are the absolute values of opposite integers always equal? Explain why or why not.

_____

_____

_____

Practice

Find each absolute value. Use the number line if needed.

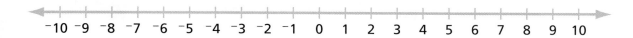

1. $|{}^-4| = 4$

2. $|0| =$

3. $|6| =$

4. $|3| =$

5. $|{}^-9| =$

6. $|8| =$

7. $|{}^-7| =$

8. $|10| =$

9. $|{}^-4| =$

10. $|{}^-10| =$

11. $|{}^-2| =$

12. $|{}^-5| =$

Find the value of each side of the equation.

13. $|4| = |{}^-4|$

    $\underline{\quad 4 \quad} = \underline{\quad 4 \quad}$

14. $|12| = |{}^-12|$

    $\underline{\qquad} = \underline{\qquad}$

15. $|50| = |{}^-50|$

    $\underline{\qquad} = \underline{\qquad}$

16. $|9| = |{}^-9|$

    $\underline{\qquad} = \underline{\qquad}$

17. $|28| = |{}^-28|$

    $\underline{\qquad} = \underline{\qquad}$

18. $|150| = |{}^-150|$

    $\underline{\qquad} = \underline{\qquad}$

Is each equation true? Write _Yes_ or _No_.

19. $|2| = |{}^-2|$ Yes

20. $|{}^-8| = |{}^-8|$

21. $|{}^-6| = |{}^-7|$

22. $|10| = |9|$

23. $|1| = |1|$

24. $|{}^-3| = |3|$

25. $|{}^-4| = |4|$

26. $|{}^-10| = |{}^-10|$

27. $|9| = |{}^-8|$

 What is the absolute value of $|{}^-12,589|$? Explain how you can find the solution.

_____

_____

_____

_____

Lesson 3

Integers and Temperature

A thermometer shows temperatures greater than and less than zero degrees in the same way a number line shows integers greater and less than zero.

Example:

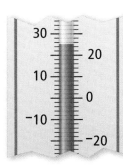

The thermometer shows a temperature of 25° F, read *twenty-five degrees Fahrenheit.*

Model 1

Is the temperature shown greater than

or less than 0°? _____

Is the temperature a positive or negative

integer? _____

How many units from 0 is the

temperature? _____

What is the temperature? _____

Fahrenheit (F)

Temperatures can be related by how many degrees there are between them.

Model 2

20° and ⁻10° have _____ degrees between them.

10° and ⁻10° have _____ degrees between them.

⁻18 and ⁻4 have _____ degrees between them.

Draw two thermometers on a separate piece of paper. On the first thermometer, show a temperature of 18°F. On the second thermometer, show a temperature of ⁻7°F.

Practice

Write each temperature shown on the thermometer.

1. Fahrenheit (F)

−20°F

2. Fahrenheit (F)

3. Fahrenheit (F)

4. Fahrenheit (F)

5. Fahrenheit (F)

6. Fahrenheit (F)

How many degrees are between each pair of temperatures?

7. −15° and −10° __5__ degrees	8. 20° and 25° _____ degrees	9. −5° and 10° _____ degrees
10. −15° and 0° _____ degrees	11. 0° and 10° _____ degrees	12. −5° and 5° _____ degrees

 Explain how you can find the distance between a negative temperature and a positive temperature.

_____

_____

_____

_____

Integer Order

Several symbols are used to compare numbers.

Examples: $<$ $>$ $=$
                 less than   greater than   equal to

You can use a number line to compare and order integers. The farther to the right an integer is on a number line, the greater its value.

Example:

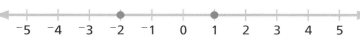

1 is to the right of $^-2$, so 1 is greater than $^-2$.
$1 > {}^-2$

Use the number line to compare integers. Write $<$, $>$, or $=$.

Model 1

$^-5$ $^-4$ $^-3$ $^-2$ $^-1$ 0 1 2 3 4 5

Graph $^-1$ and $^-4$ on the number line.	**Graph 3 and $^-3$ on the number line.**
Is $^-1$ to the left or right of $^-4$?	Is $^-3$ to the left or right of 3?
_____	_____
Is $^-1$ greater than or less than $^-4$?	Is $^-3$ greater than or less than 3?
_____	_____
$^-1$ _____ $^-4$	$^-3$ _____ 3

Write 4, $^-2$, $^-5$, and 0 in order from greatest to least.

Model 2

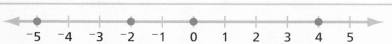

Numbers decrease in value on a number line from right to left.

4, _____, _____, $^-5$

Explain why a positive integer is always greater than a negative integer.

_____

_____

Practice

Compare each pair of integers. Write $<$, $>$, or $=$. Use the number line if needed.

-10 -9 -8 -7 -6 -5 -4 -3 -2 -1 0 1 2 3 4 5 6 7 8 9 10

1. 4 __>__ ⁻1

2. ⁻7 _____ ⁻8

3. ⁻9 _____ 6

4. ⁻5 _____ ⁻5

5. ⁻4 _____ ⁻1

6. ⁻1 _____ 0

7. ⁻2 _____ ⁻4

8. ⁻3 _____ 2

9. ⁻10 _____ ⁻6

Write each set of integers in order from least to greatest.

10. ⁻7, 3, ⁻2, 1   **⁻7, ⁻2, 1, 3**

11. ⁻4, ⁻8, 5, 0

12. ⁻9, 9, ⁻10, 0

13. ⁻5, 1, ⁻3, ⁻6

Use the number line above to answer each question.

14. What integers are between ⁻5 and ⁻2?

   ⁻4, ⁻3

15. What integers are greater than ⁻7 but less than ⁻3?

16. What integer is greater than ⁻1 and less than 1?

17. What integers are between 7 and 10?

18. What is the greatest negative integer?

19. What is the least positive integer?

 Is the opposite of an integer always less than the integer? Explain your answer.

_____

_____

_____

_____

Lesson 5

Integers and Counters

Counters can be used to model positive and negative integers. Suppose a yellow counter has a value of $^+1$ and a red counter has a value of $^-1$.

Examples:

The positive counters show 3.
The negative counters show $^-2$.

Use counters to show the integers.

Model 1 What number do the counters show?

Four positive counters

represent _____.

Six negative counters

represent _____.

One positive and one negative counter combined have a value of 0.
$1 + {}^-1 = 0$. This is called a **zero pair**.

What is the result after the zero pairs are removed?

Model 2

The result is _____.

The result is _____.

 Draw counters to show two integers that have the same absolute value. Write + and − on the counters to represent positive and negative.

Practice

What integer is represented by each set of counters?

1.

    ⁻2

2.

3.

Draw a set of counters to represent each of the following numbers.

4. 7

5. ⁻4

6. ⁻6

Find the value of each set of counters.

7.

    ⁻2

8.

9.

10.

 Explain why three positive counters and five negative counters have a value of ⁻2.

_____

_____

6 Integers and Addition

Use a number line to model addition of integers.

Example: When you add a positive integer, move to the right on the number line.

$$5 + 2 = 7$$

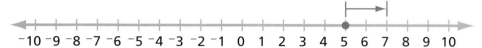

$$\begin{array}{cccccccccccccccccccccc} & & & & & & & & & & & & & & & & & & & & & \\ ^-10 & ^-9 & ^-8 & ^-7 & ^-6 & ^-5 & ^-4 & ^-3 & ^-2 & ^-1 & 0 & 1 & 2 & 3 & 4 & 5 & 6 & 7 & 8 & 9 & 10 \end{array}$$

When you add a negative integer, move to the left on the number line. Use the number line above to add the following integers.

Model 1

Add 6 + (⁻3).	Add ⁻2 + (⁻4).
From 6 on the number line, move 3 spaces to the _____.	From ⁻2 on the number line, move 4 spaces to the _____.
Stop at _____.	Stop at _____.
6 + (⁻3) = _____	⁻2 + (⁻4) = _____

Counters can be used to model addition of integers. The sum is the integer represented when all the zero pairs are removed.

Model 2

Add ⁻5 + 4.	Add ⁻3 + (⁻4).

_____ zero pairs	_____ zero pairs
_____ more negative counter	_____ negative counters
⁻5 + 4 = _____	⁻3 + (⁻4) = _____

 Describe how you could use raisins and peanuts to model adding 5 + (⁻2).

_____

_____

_____

Practice

Use the number line to add.

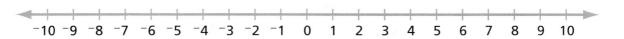

$^-10$ $^-9$ $^-8$ $^-7$ $^-6$ $^-5$ $^-4$ $^-3$ $^-2$ $^-1$ 0 1 2 3 4 5 6 7 8 9 10

1. $^-6 + 7 =$ 1

2. $8 + (^-4) =$

3. $^-5 + (^-5) =$

4. $2 + (^-9) =$

5. $^-6 + 6 =$

6. $9 + (^-5) =$

7. $^-3 + (^-7) =$

8. $5 + (^-8) =$

9. $1 + 9 =$

10. $^-4 + 0 =$

11. $8 + (^-3) =$

12. $10 + (^-12) =$

Give the sum represented by each set of counters.

13.  $^-3$

14.

15.

16.

Draw a set of counters to model each problem. Then solve.

17. $3 + (^-5) = ^-2$

18. $^-1 + 6 =$

 Draw counters to model $5 + (^-2) + (^-3)$. Give the result.

Integers and Subtraction

Remember that subtraction is the opposite of addition. Move to the left on the number line when subtracting a positive integer. Move to the right when subtracting a negative integer.

Examples: $^-1 - 3 = ^-4$

$1 - (^-3) = 4$

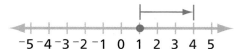

Use the number line to subtract.

Model 1 ▶ **Subtract a positive number.**

4 − 6	$^-3 - 4$

Mark a point at 4.

Mark a point at $^-3$.

What direction will you move?

What direction will you move?

_____

_____

4 − 6 = _____

$^-3 - 4$ = _____

Model 2 ▶ **Subtract a negative number.**

$4 - (^-1)$	$^-3 - (^-2)$

Mark a point at 4.

Mark a point at $^-3$.

What direction will you move?

What direction will you move?

_____

_____

$4 - (^-1)$ = _____

$^-3 - (^-2)$ = _____

 Draw a number line to model $^-4 - 2$.

Practice

Refer to the number line to subtract.

$$\xleftarrow{\qquad} \overset{-5}{|} \ \overset{-4}{|} \ \overset{-3}{|} \ \overset{-2}{|} \ \overset{-1}{|} \ \overset{0}{|} \ \overset{1}{|} \ \overset{2}{|} \ \overset{3}{|} \ \overset{4}{|} \ \overset{5}{|} \xrightarrow{\qquad}$$

1. $4 - (^-1) = 5$

2. $3 - 2 =$

3. $3 - (^-1) =$

4. $0 - 4 =$

5. $^-1 - 2 =$

6. $4 - 2 =$

7. $4 - 5 =$

8. $^-4 - (^-2) =$

9. $^-1 - 0 =$

10. $5 - 5 =$

11. $^-2 - 1 =$

12. $0 - 3 =$

Draw a number line to solve each problem.

13. $^-6 - {^-3} =$

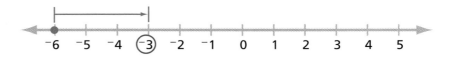

14. $2 - 7 =$

15. $^-5 - (^-8) =$

When subtracting a positive integer from a negative integer, will the difference always be a negative integer? Explain.

_____

_____

_____

Addition and Subtraction of Integers

Use these rules for adding integers.

- When the signs of both integers are positive, the sum is positive.

  Example: $2 + 5 = 7$

- When the signs of both integers are negative, the sum is negative.

  Example: $^-2 + (^-5) = ^-7$

- When the signs are different, subtract the absolute values. Then give the sum the same sign as the number with the greater absolute value.

  Examples: $^-2 + 5 = 3$ $2 + ^-5 = ^-3$

Model 1

Add. $^-5 + (^-6) = \underline{\hspace{1cm}}$	**Think:** Both integers are negative. The sum is negative.
Add. $^-9 + 3 = ?$ $\|^-9\| - \|3\| = ?$ $\underline{\hspace{1cm}} - \underline{\hspace{1cm}} = \underline{\hspace{1cm}}$ $^-9 + 3 = \underline{\hspace{1cm}}$	**Think:** The signs are different. Subtract the absolute values. Give the answer the same sign as the greater absolute value.

Use the rules for adding integers to subtract integers. Subtracting an integer is the same as adding its opposite.

Examples: $5 - 3 = 2$ means the same as $5 + (^-3) = 2$.
 $5 - (^-3) = 8$ means the same as $5 + 3 = 8$.
 $^-5 - 3 = ^-8$ means the same as $^-5 + (^-3) = ^-8$.

Model 2

Subtract. $7 - (^-2) = ?$ $7 + 2 = \underline{\hspace{1cm}}$	**Think:** Subtract by adding the opposite $+ (^+2)$.
Subtract. $^-6 - 1 = ?$ $\underline{\hspace{1cm}} + \underline{\hspace{1cm}} = \underline{\hspace{1cm}}$	**Think:** Subtract by adding the opposite $+ (^-1)$.

 How would you find the sign of the sum $9 + ^-5 + ^-9$?

_____

_____

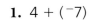

Practice

State whether the sum will be positive, negative, or zero.

1. $4 + (^-7)$ negative

2. $4 + 4$

3. $^-8 + 8$

4. $10 + (^-5)$

5. $^-9 + (^-8)$

6. $^-6 + 2$

Find the sum.

7. $2 + (^-7) = {}^-5$

8. $9 + (^-1) =$

9. $^-4 + 1 =$

10. $^-14 + 3 =$

11. $^-3 + (^-3) =$

12. $^-7 + 0 =$

13. $^-8 + (^-12) =$

14. $3 + (^-6) =$

15. $0 + (^-8) =$

16. $^-6 + 14 =$

17. $^-1 + 0 =$

18. $^-6 + 3 =$

Rewrite each subtraction problem as an addition problem. Solve.

19. $7 - (^-1) =$

$7 + 1 = 8$

20. $^-8 - 2 =$

21. $17 - (^-3) =$

22. $6 - 4 =$

23. $0 - (^-5) =$

24. $12 - (^-7) =$

Subtract.

25. $9 - (^-3) = 12$

26. $^-2 - 10 =$

27. $^-6 - 0 =$

28. $1 - 7 =$

29. $^-8 - (^-3) =$

30. $15 - (^-14) =$

 If you add four negative integers, will the sum be positive or negative? How do you know?

_____

_____

9 Equations with Integers

Solving an equation means to find the value of the variable that makes the equation true. To solve an equation, add or subtract the same number on both sides of the equal sign to show the variable by itself.

Examples:

$$x + 4 = 0 \qquad\qquad x - 2 = 5$$
$$x + 4 - 4 = 0 - 4 \qquad\qquad x - 2 + 2 = 5 + 2$$
$$x = {}^-4 \qquad\qquad\qquad x = 7$$

Solve the equation.

Model 1

$$x - 8 = {}^-2$$

$$x - 8 + \rule{2cm}{0.4pt} = {}^-2 + \rule{2cm}{0.4pt}$$

$$x = \rule{2cm}{0.4pt}$$

$$x - 8 = {}^-2$$

$$\rule{2cm}{0.4pt} - 8 = {}^-2$$

$$\rule{2cm}{0.4pt} = \rule{2cm}{0.4pt}$$

Think: The value of x should be positive.

You can check your work by substituting the value of the variable into the original equation.

Write and solve an equation to model the situation.

Model 2

The low temperature one day was $^-5°F$. The high temperature that day was $13°F$. How much did the temperature change? Let x equal the temperature change.

low temperature + temperature change = high temperature

$$\rule{2cm}{0.4pt} \quad + \quad \rule{2cm}{0.4pt} \quad = \quad \rule{2cm}{0.4pt}$$

Solve the equation. $\qquad {}^-5 + x = 13$

$$^-5 + x + \rule{2cm}{0.4pt} = 13 + \rule{2cm}{0.4pt}$$

$$x = \rule{2cm}{0.4pt}$$

The temperature changed \rule{2cm}{0.4pt} degrees.

How else could you model and solve the temperature problem using a thermometer?

_____

Practice

Add or subtract to find the value of *x*. Show each step. Use the number line if needed.

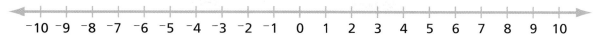

$$^-10 \quad ^-9 \quad ^-8 \quad ^-7 \quad ^-6 \quad ^-5 \quad ^-4 \quad ^-3 \quad ^-2 \quad ^-1 \quad 0 \quad 1 \quad 2 \quad 3 \quad 4 \quad 5 \quad 6 \quad 7 \quad 8 \quad 9 \quad 10$$

1. $x + 5 = 8$

$x + 5 - 5 = 8 - 5$

$x = 3$

2. $x + 3 = 3$

3. $x - 2 = {}^-6$

4. $x + ({}^-1) = 5$

5. $x - 6 = {}^-2$

6. ${}^-7 + x = {}^-7$

7. $x + 4 = {}^-6$

8. $1 + x = 0$

9. $x - 2 = {}^-5$

10. $x + 4 = {}^-5$

11. $x + 10 = {}^-10$

12. ${}^-5 + x = {}^-9$

Write and solve an equation for each situation.

13. Snack machines at one factory are located 6 floors above ground. James works 1 floor below ground. How many floors above where James works are the snack machines?

$^-1 + x = 6$

$^-1 + x + 1 = 6 + 1$

$x = 7$ floors above James

14. A day care center is on the fifth floor of an office building. It is 3 floors below a gift shop. What floor is the gift shop on?

Write and solve a word problem that uses adding or subtracting integers.

_____

_____

_____

Strength Builder

▶ Absolute Value Match

Two or more players can play *Absolute Value Match*. Players try to remember where pairs of numbers with the same absolute value are located.

Materials
- 20 plain index cards
- pencil

Getting Ready
1. Write the numbers from $^-10$ through $^-1$, and 1 through 10 on index cards.
2. Shuffle the integer cards and place them facedown on the table in four rows of five cards each.

● ●

Playing the Game
1. The first player turns over two cards. If the cards have the same absolute value, the player keeps the pair. Then he or she may take another turn. The player's turn stops when two cards do not have the same absolute value.
2. If the cards do not have the same absolute value, they are placed back where they were. Then the next player takes a turn.
3. Players continue to take turns until all the pairs are found.
4. When all the cards are gone, the player with the most pairs wins.

Other Ways to Play
1. You may want to have each player turn over only two cards on each turn. Even if a match is made, a player does not get another turn until the next round.
2. The game can also be played with a timer. Give a player one minute to find as many pairs as he or she can. Shuffle the cards, place them flat again, and let the next player find pairs for one minute. The winner is the player with the greatest number of pairs found in a minute.

▶ Zero Pairs

Two or more players can play *Zero Pairs*. Players try to get rid of zero pairs of numbers until they are out of cards.

Materials

- two sets of integer cards from the *Absolute Value Match* game

Getting Ready

1. Shuffle two sets of cards and deal 5 cards to each player.
2. Place the stack of cards facedown on the table.

Playing the Game

1. The first player looks for any zero pairs in his or her hand. If the player has a zero pair, the zero pair can be placed in the discard pile. That player would not have to take any cards.
2. If the player does not have a zero pair, he or she must draw two cards from the deck. Then it is the next player's turn.

3. Players continue to take turns. If you run out of cards, shuffle the discard pile.
4. Play continues until someone gets rid of all cards in his or her hand. When that happens, the player says, "Zero Pairs." The first player to run out of cards wins.

▶ Adding Integers

Two or more players can play *Adding Integers.* Players gain points by adding two integers at a time.

Materials

- one set of integer cards from the *Absolute Value Match* game
- paper and pencil to write problems and record scores

Getting Ready

1. Shuffle the set of integer cards.
2. Place the stack of cards facedown on the table.

Playing the Game

1. Player one draws two cards from the deck and adds them. He or she may use paper or mental math to add the numbers. The other players check the answer. If the answer is correct, the sum is the player's score for the round. If the answer is not correct, the player gets no points.

2. Other players take turns drawing cards and adding to find their scores for each round.
3. Cards that have been used are placed in a discard pile. When you run out of cards, shuffle the discard pile.
4. Play continues for five rounds. Then players add their scores. The player with the greatest number of points wins the game.

Match each point to its value.

A B   C D E   F G H   I J K   L M
-10 -9 -8 -7 -6 -5 -4 -3 -2 -1 0 1 2 3 4 5 6 7 8 9 10

1. ⁻4	**2.** 5
3. |⁻6|	**4.** the opposite of 1
5. the opposite of ⁻8	**6.** 0
7. |⁻4|	**8.** |9|

Write each temperature.

9.

10.

11.
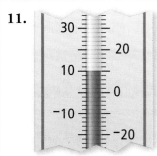

Write in order from least to greatest.

12. ⁻8, 9, ⁻5, 2	**13.** ⁻6, 10, 0, ⁻2

Add or subtract.

14. ⁻5 + 2 =	**15.** ⁻6 − 4 =	**16.** ⁻4 + ⁻4 =
17. 3 − 5 =	**18.** ⁻5 + ⁻5 =	**19.** ⁻1 − 0 =
20. ⁻2 + 8 =	**21.** 6 − 7 =	**22.** 5 + 3 =

Find the value of *x*. Show each step.

23. 3 + x = 2	**24.** x − 5 = 5	**25.** ⁻1 + x = ⁻7

Cumulative Review

Write an expression to model each situation. Then find the value of each expression.

1. John is 4 years younger than his sister. His sister is 17 years old.

   Expression:

   Value:

2. Susana had 16 stuffed animals. She gave 3 away and got one new stuffed animal.

   Expression:

   Value:

Substitute the given value for the variable. Write the value of each expression.

3. If $n = 8$, what is $n - 2$?

   The value is _____.

4. If $n = 50$, what is $n + 27$?

   The value is _____.

Solve each equation. Check your answer.

5. $8 \times n = 88$

   $n =$

6. $32 + n = 35$

   $n =$

7. $48 - n = 36$

   $n =$

8. $14 \times 2 = 2 \times n$

   $n =$

9. $3 \times n = 0$

   $n =$

10. $(5 + n) + 6 = 5 + (1 + 6)$

    $n =$

11. $27 + 0 = n$

    $n =$

12. $5 = 5 \times n$

    $n =$

13. $6 \cdot (3 \cdot 10) = 6 \cdot (n \cdot 3)$

    $n =$

Solve each equation. Name the addition or multiplication property you used.

14. $5 + 2 = 2 + n$

    $n =$

15. $4 \times n \times 5 = 0$

    $n =$

16. $(9 + 5) + 1 = n + (5 + 1)$

    $n =$

Find the value of each expression.

17. $70 \div 2(6 + 1)$

18. $(3 + 2) \times 8 + 2$

19. $7 + 3^2 - 5$

Cumulative Review

20. Graph and label the points of the function.

Input x	Output y
2	0
4	2
6	4
8	6

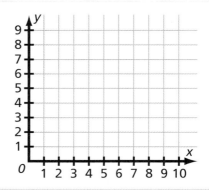

Determine if the ordered pair is a solution of the given equation.

21. $(1, 4)$ $y = x - 3$

22. $(4, 4)$ $y = x$

23. $(3, 5)$ $y = 3x - 4$

Complete the table for the equation $y = 2x$. Graph the solutions and label the points. Then draw the line of the equation.

	Input x	Linear equation y = 2x	Output y	Ordered Pair (x, y)
24.	1	$y =$		
25.	2	$y =$		
26.	3	$y =$		

27.

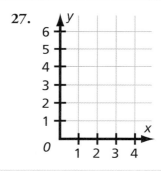

Add or subtract as indicated.

28. $3 + {}^-9 =$

29. $^-8 - 7 =$

30. $^-12 + 7 =$

31. $^-6 - 1 =$

32. $15 + {}^-17 =$

33. $20 - ({}^-4) =$

Find the value of x.

34. $4 - x = 5$

$x =$

35. $^-2 + x = {}^-8$

$x =$

36. $x - 6 = {}^-10$

$x =$

37. $x + {}^-7 = {}^-1$

$x =$

Glossary

Absolute value (p. 58) An integer's distance from 0 on the number line; the symbol for absolute value is | |

$|5| = 5$ *and* $|^-5| = 5$

Associative Property of Addition (p. 22) States that the grouping of addends does not change the sum of those addends

$(6 + 2) + 1 = 6 + (2 + 1)$

Associative Property of Multiplication (p. 26) States that changing the grouping of factors does not change the product of those factors

$(2 \times 1) \times 4 = 2 \times (1 \times 4)$

Balanced (p. 14) Equations that have the same value on both sides of the equal sign

Commutative Property of Addition (p. 20) States that changing the order of addends does not change the sum

$3 + 7 = 7 + 3$

Commutative Property of Multiplication (p. 24) States that changing the order of factors does not change the product of those factors

$3 \times 4 = 4 \times 3$

Coordinate grid (p. 44) A graph formed by intersecting and perpendicular lines, marked by an *x*-axis and a *y*-axis

Distributive Property of Multiplication (p. 30) States that multiplying a sum by a number is the same as multiplying each addend by the number and then adding the product

$2(10 + 3) = 2 \cdot 10 + 2 \cdot 3$

Equation (p. 8) A mathematical sentence that has an equal sign

$x + 2 = 5$

Expression (p. 4) A mathematical phrase that does not have an equal sign

$4 + 3$

Function (p. 42) Rule that describes the relationship between input and output values

Function table (p. 42) Arrangement in rows and columns showing both input and output values for a given function

Identity Property of Multiplication (p. 28) States that when 1 is multiplied by a number, the product is that number

$5 \times 1 = 5$

Integers (p. 56) Whole numbers and their opposites

$\{^-3, ^-2, ^-1, 0, 1, 2, 3\}$

Line graph (p. 40) A type of graph that shows change over a period of time

Glossary

Linear equation (p. 50) An equation whose solution lies on a line in a graph

$y = x + 5$

Opposites (p. 56) Numbers that are the same distance from zero on the number line, but in opposite directions

The opposite of 2 is $^-2$.

Order of Operations (p. 32) A standard order for finding the value of a math sentence: parentheses, exponents, multiplication and division, addition and subtraction

Ordered pair (p. 44) Gives the exact location of a point on a grid by the x-value and y-value

Parentheses (p. 22) Symbols used to group numbers to indicate what operations are to be done first

Property of One (for multiplication) (p. 28) States that when 1 is multiplied by a number, the product is that number

$5 \times 1 = 5$

Solution (p. 12) The value of the variable that makes an equation true

Solution set (p. 46) All the possible values of the variables that make an equation true

Substitution (p. 12) To replace a variable with a specific amount to solve an equation

Table (p. 38) A way to organize data in rows and columns so that it is easy to read

Thermometer (p. 60) Shows temperatures in degrees

Value (p. 4) Numerical amount

The value of $28 \div 2$ is 14.

Variable (p. 6) A letter or symbol that stands for a number

Zero pair (p. 64) Two integers whose sum is zero

4 and $^-4$

Zero Property of Addition (p. 20) States that the sum of zero and a number is that same number

$3 + 0 = 3$

Zero Property of Multiplication (p. 28) States that when zero is a factor, the product is zero

$5 \times 0 = 0$